Saving Our Planet

Saving Earth's Land

by Brienna Rossiter

FOCUS READERS®
PIONEER

www.focusreaders.com

Focus Readers is distributed by North Star Editions:
sales@northstareditions.com | 888-417-0195

Produced for Focus Readers by Red Line Editorial.

Photographs ©: Shutterstock Images, cover, 1, 4, 7, 8, 11, 12, 15, 16, 19 (top), 19 (bottom), 21

Library of Congress Cataloging-in-Publication Data
Names: Rossiter, Brienna, author.
Title: Saving earth's land / by Brienna Rossiter.
Description: Lake Elmo, MN : Focus Readers, [2022] | Series: Saving our planet | Includes index. | Audience: Grades 2-3
Identifiers: LCCN 2021006313 (print) | LCCN 2021006314 (ebook) | ISBN 9781644938232 (hardcover) | ISBN 9781644938690 (paperback) | ISBN 9781644939154 (ebook) | ISBN 9781644939604 (pdf)
Subjects: LCSH: Environmental sciences--Juvenile literature. | Environmentalism--Juvenile literature. | Green movement--Juvenile literature.
Classification: LCC GE105 .R67 2022 (print) | LCC GE105 (ebook) | DDC 363.739/6--dc23
LC record available at https://lccn.loc.gov/2021006313
LC ebook record available at https://lccn.loc.gov/2021006314

Printed in the United States of America
Mankato, MN
082021

About the Author

Brienna Rossiter is a writer and editor who lives in Minnesota. She loves hiking in the woods.

Table of Contents

Chapter 1

Using Land

People use land in many ways. For example, people cut down trees to get wood. They dig **minerals** from the ground. They also use land to raise animals and crops.

But people must be careful. Some ways of using land are harmful. They can cause damage or **pollution**.

Also, many **resources** are not **renewable**. They will run out if people use too much.

Natural Resources

Some of Earth's resources are renewable. Others are not.

coal
(not renewable)

oil
(not renewable)

sunlight
(renewable)

trees
(renewable)

water
(renewable)

wind
(renewable)

Chapter 2

Harmful Actions

Some human actions are bad for Earth. For example, people cut down too many trees. They often replace forests with buildings or farms.

Also, cities, farms, and factories make waste. Much of this waste goes to **landfills**. But some of it leaks out. Other waste is dumped or spilled. It can get into the ground.

Did You Know?

Waste can be paper, plastic, **chemicals**, and more.

Chapter 3

Problems

Earth needs many kinds of plants and animals to stay healthy. But people have damaged many **habitats**. Some plants and animals can no longer live there. They are in danger of dying out.

Landfills take up a lot of space. And waste takes a long time to break down. It can also release harmful chemicals. They can make people or animals sick.

Did You Know?

Plastic takes hundreds of years to break down.

Chapter 4

What We Can Do

People can help take care of Earth's land. They can clean up waste. They can **restore** damaged areas. And they can plan for the future.

Cities and countries can make rules. They can limit the resources people use. They can set aside natural areas. And they can plan safe ways to store waste. These actions will help keep Earth clean and beautiful.

Recycle

Make Less Waste

You can help make less waste. Buy only what you need. And look for ways to use things again. For example, bring your own bags when you shop. You can also **recycle**. Many items can go in a recycling bin. People can use the materials to make new things.

FOCUS ON

Saving Earth's Land

Write your answers on a separate piece of paper.

1. Write a sentence describing the main ideas of Chapter 1.
2. Do you think it's more important to fix land that was damaged in the past or to stop future damage? Why?
3. Where does the waste in landfills come from?
   A. wild animals
   B. cities and factories
   C. forests and natural areas
4. What is an example of creating less waste?
   A. throwing away a piece of paper
   B. using a shopping bag many times
   C. buying lots of plastic bottles

Answer key on page 24.

Glossary

chemicals
Basic forms of matter. Some chemicals are helpful, and some are harmful.

habitats
Places where animals live.

landfills
Places where garbage is sent to be stored.

minerals
Solid materials that are found in nature.

pollution
Harmful chemicals that get into water, air, or soil.

recycle
To use something again, in a different form or for a different purpose.

renewable
Possible to get or make more of.

resources
Parts of nature that are useful to humans. Resources include sunlight, water, minerals, and plants.

restore
To bring something back to the way it used to be.

To Learn More

BOOKS

Haley, Charly. *Earth Day*. Minneapolis: Abdo Publishing, 2019.

Huddleston, Emma. *Planting a Seedling*. Lake Elmo, MN: Focus Readers, 2021.

NOTE TO EDUCATORS

Visit **www.focusreaders.com** to find lesson plans, activities, links, and other resources related to this title.

Index

Answer Key: 1. Answers will vary; **2.** Answers will vary; **3.** B; **4.** B